Study Guide for Human Anatomy and Physiology:
Lymphatic System, Immunity, Respiratory System and Digestive System

TERMINOLOGY

A lack of resistance in human beings is called:
a. non-resistance
b. susceptibility
c. immunocompromised
c. immunocompetent

Answer: B

There are _____ general types of immunity.
a. one
b. two
c. three
d. four

Answer: B

_____ immunity is defined as the body's response to a wide range of pathogens (disease causing organisms).
a. Adaptive
b. Humoral
c. Cell-mediated
d. Innate

Answer: D

_____ immunity is defined as the stimulation and activation of specific lymphocytes to fight a specific foreign substance.
a. Innate
b. Adaptive
c. Herd
d. Communal

Answer: B

The ability of human beings to fight off disease is referred to as _____.
a. resistance
b. competency
c. immunocompetency
d. susceptibility

Answer: A

STRUCTURE AND FUNCTION

The function of the lymphatic system is to:
a. drain interstitial fluid
b. protect the body against invasion
c. transport dietary fats
d. A, B and C are correct

Answer: D

The fluid that flows through the lymphatic system is called:
a. lymph
b. interstitial fluid
c. dietary fats
d. blood

Answer: A

Select the correct sequence regarding lymph flow.
a. Lymph capillaries, lymphatic vessels, lymph trunks, thoracic duct and right lymphatic duct, subclavian veins
b. Subclavian veins, thoracic duct and right lymphatic duct, lymph trunks, lymphatic vessels, lymph capillaries
c. Blood capillaries, lymph capillaries, lymphatic vessels, lymph trunks, thoracic duct and right lymphatic duct, vena cavae
d. Aorta, Arteries, Arterioles, Capillaries, Veins, Venules, Vena cavae

Answer: A

Select the mechanisms that aid in the flow of lymph.
a. Valves in lymphatic vessels
b. Respiratory movements
c. Skeletal muscle contactions
d. A, B and C are correct

Answer: D

T cells mature in the _____.
a. thyroid gland
b. thymus gland
c. red bone marrow
d. lymph nodes

Answer: B

Lymph enters lymph nodes through _____ and exits through _____.
a. efferent lymphatic vessels; afferent lymphatic vessels
b. lymph trunks; lymph ducts
c. afferent lymphatic vessels; efferent lymphatic vessels
d. lymphatic vessels; lymph trunks

Answer: C

B lymphocytes and T lymphocytes proliferate in (the):
a. the spleen
b. thymus
c. lymphatic nodules
d. lymph nodes

Answer: D

The functions of the spleen include:
a. the destruction of blood-borne pathogens
b. the destruction of worn-out red blood cells via phagocytosis
c. to carry out immune functions
d. A, B and C are correct

Answer: D

Select the incorrect statement.
a. Lymphatic nodules are found throughout the mucosa of the GI and respiratory tracts.
b. Lymphatic nodules are also called MALT.
c. Lymphatic nodules are found throughout the mucosa of the urinary and reproductive tracts.
d. Lymphatic nodules are also called lymph nodes.

Answer: D

Select the term that does not belong.
a. Lymph nodes
b. Spleen
c. Lymphatic nodules
d. Secondary lymphatic organs
e. Thymus

Answer: E

Select the term that does not belong.
a. Spleen
b. Red bone marrow
c. Thymus

d. Primary lymphatic organs

Answer: A

INNATE IMMUNITY

Innate immunity involves the:
a. first line of defense
b. second line of defense
c. third line of defense
d. A and B are correct
e. A, B and C are correct

Answer: D

The purpose of inflammation is to:
a. enhance the effects of fever
b. stimulate the release of acidic substances
c. dispose of microbes, toxins or foreign substances at injury sites
d. prepare the site of injury for repair
e. C and D are correct

Answer: E

The first line of defense includes:
a. internal defenses
b. chemical factors
c. physical factors
d. B and C are correct
e. A and B are correct

Answer: D

Which of the following is not a physical factor involved in the first line of defense?
a. Epidermis and mucous membranes
b. Mucus, hairs and cilia
c. Lacrimal apparatus, saliva and urine
d. Sebum and lysozyme

Answer: D

Which of the following is not a chemical factor involved in the first line of defense?
a. Gastric juice and lysozyme
b. Vaginal secretions
c. Defecation and vomiting
d. Sebum

Answer: C

The advantages of a fever include:
a. enhances the effects of interferons
b. inhibits the growth of some microorganisms
c. accelerates the body reactions that aid in repair
d. A, B and C are all correct

Answer: D

Which of the following is not an antimicrobial substance involved in the second line of defense?
a. Iron-binding proteins
b. Complement
c. Interferons
d. Antimicrobial proteins
e. None of these choices

Answer: E

Select the substances released by natural killer cells that are used to kill their target cells.
a. Mucinase
b. Perforin
c. Granzymes
d. Lysozymes
e. B and C are correct

Answer: E

ADAPTIVE IMMUNITY

Adaptive immunity involves:
a. eosinophils
b. basophils
c. lymphocytes
d. erythrocytes

Answer: C

B cells or B lymphocytes originate in the:
a. thymus
b. spleen
c. thyroid
d. red bone marrow

Answer: D

T cells or T lymphocytes originate in the:
a. thymus
b. lymph nodes
c. red bone marrow
d. yellow bone marrow

Answer: C

B cells mature in _____ and T cells mature in the _____.
a. thymus; thymus
b. red bone marrow; thymus
c. red bone marrow; red bone marrow
d. spleen; thymus

Answer: B

The mature T cells that exit the thymus gland are called:
a. CD4 T cells
b. CD8 T cells
c. helper T cells
d. cytotoxic T cells
e. A, B, C and D are all correct

Answer: E

Before T cells exit the thymus or B cells exit the red bone marrow, they develop _____.
a. resistance
b. specificity
c. immunocompetence
d. competence

Answer: C

_____ is the process in which a lymphocyte proliferates and differentiates in response to a specific antigen.
a. Immunity
b. Clonal selection
c. Immunocompetence
d. Memory

Answer: B

_____ or Ags are chemical substances that are recognizes as being foreign to your immune system.
a. Antibodies

b. Antigens
c. Antigen-antibody complexes
d. Cytokines

Answer: B

Which of the following is not an antigen presenting cell?
a. Eosinophil
b. Macrophage
c. B cell
d. Dendritic cell

Answer: A

All cells except _____ display major histocompatibility complexes (type I) on their surfaces.
a. B cells
b. T cells
c. RBCs
d. Platelets

Answer: C

Exogenous antigens are formed _____ body cells and endogenous antigens are formed _____ body cells.
a. outside; outside
b. inside; outside
c. outside; inside
d. inside; inside

Answer: C

CELL-MEDIATED IMMUNITY

Cell-mediated immunity begins with the activation of a low number of _____ by a specific _____.
a. T cells; antibody
b. B cells; antigen
c. T cells; antigen
d. B cells; antibody

Answer: C

Select the protein displayed by helper T cells.
a. CD2
b. CD4
c. CD8

d. CD10

Answer: B

Select the protein displayed by cytotoxic T cells.
a. CD6
b. CD8
c. CD12
d. CD24

Answer: B

Cytotoxic T cells recognize antigen fragments associated with _____ molecules.
a. MHC-II
b. MHC-III
c. MHC-I
d. MHC-IV

Answer: C

Helper T cells recognize antigen fragments associated with _____ molecules.
a. MHC-I
b. MHC-II
c. MHC-III
d. MHC-IV

Answer: B

Activation of T cells requires costimulation by _____.
a. cytokines
b. interferon
c. plasma membrane molecules
d. A and B are correct
e. A and C are correct

Answer: E

How do active cytotoxic T cells eliminate invaders?
a. Releasing perforin
b. Releasing granzymes
c. Releasing cytokines
d. Releasing antibodies
e. A and B are correct

Answer: E

ANTIBODY-MEDIATED IMMUNITY

Which of the following secretes antibodies?
a. T cells
b. B cells
c. Plasma cells
d. Macrophages

Answer: C

Which of the following is not an action of an antibody?
a. Enhancement of phagocytosis
b. Agglutination and precipitation of antigen
c. Activation of complement
d. Immobilization of bacteria
e. Neutralization of antigen
f. A, B, C, D and E are all correct

Answer: F

An antibody is a _____ that combines specifically with the antigen that triggered its production.
a. lipid
b. protein
c. nucleic acid
d. carbohydrate

Answer: B

Antibody-mediated immunity is triggered by the activation of a _____ cell by a specific antigen.
a. T
b. B
c. dendritic
d. phagocytic

Answer: B

This immunoglobulin is found on mast cells as well as basophils.
a. IgA
b. IgD
c. IgE
d. IgG
e. IgM

Answer: C

This immunoglobulin is found on the surfaces of B cells.

a. IgD
b. IgA
c. IgM
d. IgE

Answer: A

The only immunoglobulin to cross the placenta is:
a. IgG
b. IgD
c. IgE
d. IgA

Answer: A

Which of the following is found in sweat, saliva, tears, mucus, GI secretions as well as breast milk?
a. IgM
b. IgG
c. IgA
d. IgE

Answer: C

An antibody can assume a _____ shape because of the hinge regions.
a. R or D
b. Y or T
c. L or Y
d. C or O

Answer: B

The only immunoglobulin that can occur as a pentamer is:
a. IgD
b. IgE
c. IgG
d. IgA
e. IgM

Answer: E

The amount of antibody in serum is referred to as _____.
a. antibody titer
b. antigen titer
c. acute titer
d. convalescent titer

Answer: A

STRESS

Which of the following can influence your level of health and the course of disease?
a. Feelings
b. Moods
c. Beliefs
d. Thoughts
e. A, B, C and D are all correct

Answer: E

Adequate sleep and nutrition are important for a healthy _____ system.
a. respiratory
b. digestion
c. immune
d. nervous

Answer: C

Under stress:
a. individuals are less likely to eat well
b. individuals are less likely to exercise regularly
c. individuals smoke more
d. individuals consume more alcohol
e. A, B, C and D are all correct

Answer: E

ANATOMY OF THE RESPIRATORY SYSTEM

Select the following system that helps the respiratory system supply oxygen and remove carbon dioxide from the blood.
a. Endocrine system
b. Cardiovascular system
c. Lymphatic system
d. Nervous system

Answer: B

The respiratory system consists of all of the following except:
a. Nose and pharynx
b. Larynx and trachea
c. Esophagus and duodenum

d. Bronchi and lungs

Answer: C

Naso-, oro- and laryngo- are all regions of the _____.
a. nasal cavity
b. trachea
c. larynx
d. pharynx

Answer: D

The functions of the nose include _____.
a. warming air
b. moistening air
c. filtering air
d. olfaction and speech
e. A, B, C and D are all correct

Answer: E

The voice box is also known as the:
a. trachea
b. larynx
c. pharynx
d. nares

Answer: B

Which of the following is not contained in the larynx?
a. Epiglottis
b. Cricoid cartilage
c. Thyroid cartilage
d. Arytenoid, corniculate and cuneiform cartilages
e. None of these choices

Answer: E

C-shaped rings of cartilage are found extending from the larynx to the:
a. secondary bronchi
b. primary bronchi
c. tertiary bronchi
d. bronchioles
e. terminal bronchioles

Answer: B

The walls of the bronchioles contain _____.
a. increasing amounts of smooth muscle
b. decreasing amounts of cartilage
c. increasing amounts of cartilage
d. decreasing amounts of smooth muscle
e. A and B are correct

Answer: E

The lungs are paired organs enclosed by the _____.
a. abdominal cavity
b. pelvic cavity
c. pleural cavity
d. pericardial cavity

Answer: C

The throat is also known as the:
a. internal nares
b. pharynx
c. larynx
d. bronchial tree

Answer: B

The superficial layer that lines the thoracic cavity is referred to as the:
a. visceral pericardium
b. visceral pleura
c. parietal pleura
d. parietal pericardium
e. greater omentum

Answer: C

The deep layer that covers the lungs is called the _____.
a. visceral pleura
b. visceral pericardium
c. parietal pleura
d. parietal pericardium
e. lesser omentum

Answer: A

The anterior portion of the nasal cavity is called the _____.
a. internal nares

b. external nares
c. vestibule
d. septum

Answer: C

Select the term that does not belong.
a. Bronchial tree
b. Trachea
c. Primary bronchi
d. Secondary bronchi
e. Tertiary bronchi
f. Bronchiles
g. Terminal bronchioles
h. Larynx

Answer: H

Lung cells are also called _____.
a. bronchioles
b. alveoli
c. sacs
d. ducts

Answer: B

Gas exchange occurs across respiratory _____.
a. ducts
b. bronchioles
c. membranes
d. walls

Answer: C

The right lung has _____ lobes.
a. four
b. three
c. two
d. five

Answer: B

The left lung has _____ lobes.
a. two
b. three
c. four

d. six

Answer: A

PULMONARY VENTILATION

Pulmonary ventilation includes:
a. exhalation
b. inhalation
c. expiration
d. inspiration
e. A, B, C and D are all correct

Answer: E

According to Boyle's law, the volume of a gas varies _____.
a. directly with the pressure
b. inversely with the temperature
c. inversely with pressure
d. directly with the temperature

Answer: C

Select the incorrect statement.
a. Inhalation (or inspiration) occurs when the alveolar pressure falls below atmospheric pressure.
b. Contraction of the external intercostals muscles and the diaphragm decrease the intrapleural pressure and increase the volume of the thoracic cavity so the lungs can expand.
c. As the lungs expand, the alveolar pressure falls allowing air to move down the pressure gradient from the atmosphere into our lungs.
d. During forceful inhalation, no accessory muscles are needed.

Answer: D

The surface tension exerted by alveolar fluid is:
a. decreased by high lung compliance
b. decreased by low lung compliance
c. increased by surfactant
d. decreased by surfactant

Answer: D

Compliance is the:
a. ability of the lungs and thoracic wall to expand easily
b. ability of the lungs and thoracic wall to atrophy
c. exchange of oxygen for carbon dioxide at the pulmonary capillaries
d. exchange of oxygen for carbon dioxide at the systemic capillaries

Answer: A

These accessory muscles are used during forceful inhalation.
a. Scalenes
b. Pectoralis minors
c. Sternocleidomastoids
d. A, B and C are all correct

Answer: D

These accessory muscles are used during forceful exhalation.
a. Internal intercostals
b. Abdominal muscles
c. External intercostals
d. A and B are correct
e. B and C are correct

Answer: D

A deep, long-drawn inhalation followed by the complete closure of the rima glottidis is characteristic of _____.
a. sneezing
b. coughing
c. sighing
d. yawning
e. crying

Answer: B

Spasmodic contraction of the diaphragm followed by the spasmodic closure of the rima glottidis is characteristic of _____.
a. laughing
b. hiccupping
c. sobbing
d. yawning
e. sighing

Answer: B

LUNG VOLUMES AND LUNG CAPACITIES

Select the term that does not belong.
a. IRV
b. ERV
c. RV

d. TV
e. None of these choices

Answer: E

The IRV stands for _____.
a. inspiratory reserve volume
b. inhalation residual volume
c. inspiratory residual volume
d. inspiratory respiratory volume

Answer: A

The IRV is approximately _____ in males and _____ in females.
a. 3100 mL; 1900 mL
b. 1200 mL; 900 mL
c. 1200 mL; 1100 mL
d. 750 mL; 500 mL

Answer: A

The ERV (expiratory reserve volume) is approximately _____ in males and _____ in females.
a. 1200 mL; 700 mL
b. 1000 mL; 1200 mL
c. 1900 mL; 3100 mL
d. 1200 mL; 1100 mL

Answer: A

After the expiratory reserve volume is exhaled, the volume of air that remains to keep the alveoli slightly inflated is referred to as the _____.
a. inspiratory reserve volume
b. functional residual capacity
c. total lung capacity
d. residual volume
e. minimal volume

Answer: D

The inspiratory capacity is equal to:
a. TV + IRV
b. ERV + RV
c. VC + RV
d. IRV + TV + ERV

Answer: A

The functional residual capacity is the sum of:
a. residual volume and expiratory reserve volume
b. inspiratory reserve volume, tidal volume and expiratory reserve volume
c. vital capacity and residual volume
d. tidal volume and inspiratory reserve volume

Answer: A

The total lung capacity is approximately _____ in males and _____ in females.
a. 6000 mL; 5000 mL
b. 6000 mL; 4200 mL
c. 6000 mL; 3800 mL
d. 5000 mL; 4000 mL

Answer: B

Select the incorrect pairing.
a. Very deep breath: IRV
b. Inhale normally then exhale forcibly: ERV
c. IRV + TV + ERV: Vital capacity
d. VC + RV: Total lung capacity
e. None of these choices

Answer: E

OXYGEN AND CARBON DIOXIDE EXCHANGE

According to _____, each gas in a mixture of gases can exert its own pressure.
a. Dalton's law
b. Boyle's law
c. Charles' law
d. Henry's law

Answer: A

During internal respiration, both oxygen and carbon dioxide will diffuse from areas of:
a. lower partial pressures to areas of higher partial pressures
b. higher partial pressures to areas of lower partial pressures
c. higher partial pressures to areas of lower partial volumes
d. lower partial pressures to areas of higher partial volumes

Answer: B

During external respiration, both oxygen and carbon dioxide will diffuse from areas of _____.
a. higher partial pressures to areas of lower partial pressures

b. higher partial volumes to areas of lower partial volumes
c. lower partial volumes to areas of higher partial volumes
d. lower partial pressures to areas of higher partial volumes

Answer: A

External respiration is also called _____.
a. pulmonary gas exchange
b. internal respiration
c. systemic gas exchange
d. None of these choices

Answer: A

Internal respiration is also called _____.
a. external respiration
b. pulmonary ventilation
c. pulmonary gas exchange
d. systemic gas exchange

Answer: D

External respiration is the exchange of gases between _____.
a. alveoli and systemic blood capillaries
b. alveoli and pulmonary blood capillaries
c. systemic blood capillaries and tissue cells
d. pulmonary blood capillaries and tissue cells

Answer: B

Internal respiration is the exchange of gases between _____.
a. systemic blood capillaries and pulmonary blood capillaries
b. alveoli and tissue cells
c. systemic blood capillaries and tissue cells
d. pulmonary blood capillaries and tissue cells

Answer: C

OXYGEN AND CARBON DIOXIDE TRANSPORT

How does oxygen primarily travel through the blood (in each 100 mL of oxygenated blood)?
a. Converted to bicarbonate
b. Dissolved in plasma
c. Bound to hemoglobin as oxyhemoglobin
d. Bound to hemoglobin as carbaminohemoglobin

Answer: C

Which of the following does not affect the binding of oxygen to hemoglobin?
a. Acidity
b. Partial pressure of carbon monoxide
c. Temperature
d. Partial pressure of oxygen
e. BPG

Answer: B

Select the incorrect statement.
a. 7 % of carbon dioxide is dissolved in blood plasma in each 100 mL of deoxygenated blood
b. 23 % of carbon dioxide combines with hemoglobin as carbaminohemoglobin in each 100 mL of deoxygenated blood
c. 1.5 % of carbon dioxide is dissolved as oxygen in blood plasma in each 100 mL of deoxygenated blood
d. 70 % of carbon dioxide is converted to bicarbonate in each 100 mL of deoxygenated blood

Answer: C

CONTROL OF RESPIRATION

Which of the following structures is associated with respiratory control?
a. Midbrain
b. Medulla
c. Pons
d. Cerebrum
e. B and C are correct

Answer: E

The _____ area sets the basic rhythm of respiration.
a. expiratory
b. inspiratory
c. pneumotaxic
d. apneustic

Answer: B

_____ is the deficiency of oxygen at the tissue level.
a. Hypocapnia
b. Hypocarbia
c. Hyperventilation
d. Hypoxia

Answer: D

Which of the following factors does not contribute to the regulation of respiration?
a. Pain and temperature
b. Limbic system
c. Stretching the anal sphincter muscle
d. Irritation of airways and Blood pressure
e. None of these choices

Answer: E

These areas coordinate the transition between inhalation and exhalation.
a. Inspiratory area
b. Expiratory area
c. Penumotaxic area
d. Apneustic area
e. A and B are correct
f. C and D are correct

Answer: F

EXERCISE

When muscles contract during exercise, they produce large amounts of _____ and consume large amounts of _____.
a. oxygen; carbon dioxide
b. nitrogen; oxygen
c. carbon dioxide; nitrogen
d. carbon dioxide; oxygen

Answer: D

At the onset of exercise, there is an abrupt increase in ventilation due to:
a. neural changes that send excitatory impulses to the pneumotaxic area in the pons.
b. neural changes that send inhibitory impulses to the inspiratory area in the medulla.
c. neural changes that send excitatory impulses to the apneustic area in the pons.
d. neural changes that send excitatory impulses to the inspiratory area in the pons.

Answer: D

During moderate exercise, the more gradual increase in ventilation is due to:
a. chemical changes in the blood
b. physical changes in the blood
c. chemical changes in the lymph
d. physical changes in the urine
e. A and B are correct

Answer: E

Pulmonary perfusion is:
a. an increase in blood flow to the lungs due to increased urinary output
b. an increase in blood flow to the lungs due to increased cardiac output
c. an increase in blood flow to the lungs due to changes in the pH of the blood
d. an increase in blood flow to the lungs due to temperature fluctuations

Answer: B

At the end of exercise, the initial decrease in pulmonary ventilation is due to:
a. neural factors once movement stops or slow
b. chemical factors
c. temperature fluctuations
d. pH changes

Answer: A

At the end of exercise, the more gradual decrease in pulmonary ventilation is due to:
a. the slower return of temperature to the resting state
b. the slower return of blood chemistry levels to the resting state
c. the slower return of blood flow to the pulmonary capillaries
d. the slower return of glucose to alveolar cells
e. A and B are correct

Answer: E

OVERVIEW OF THE DIGESTIVE SYSTEM

The digestive system consists of:
a. GI tract
b. accessory organs
c. blood vessels
d. lymph nodes
e. A and B are correct

Answer: E

Select the structures that make up the GI tract.
a. Mouth, Pharynx, Esophagus
b. Stomach
c. Small intestine and Large intestine
d. Salivary glands and Liver
e. A, B and C are correct

Answer: E

Where does the digestive tract begin?
a. Mouth
b. Pharynx
c. Stomach
d. Small intestine

Answer: A

Where does the digestive tract end?
a. Small intestine
b. Liver
c. Large intestine
d. Anus

Answer: D

Select the structures that are regarded as accessory organs for this system.
a. Teeth, Tongue
b. Salivary glands
c. Liver and Pancreas
d. Heart and Kidneys
e. A, B and C are all correct

Answer: E

FUNCTION

The process of materials entering the digestive system using the mouth is referred to as _____.
a. mechanical processing
b. digestion
c. ingestion
d. absorption

Answer: C

The release of water, enzymes, acids, salts and buffers is referred to as _____.
a. secretion
b. excretion
c. digestion
d. absorption

Answer: A

_____ is the removal of waste products from body fluids.

a. Ingestion
b. Excretion
c. Secretion
d. Absorption

Answer: B

Select the functions of the digestive system.
a. Ingestion and Mechanical processing
b. Digestion and Secretion
c. Absorption and Excretion
d. A, B and C are all correct

Answer: D

MEMBRANES

The serosa or _____ covers organs in the peritoneal cavity.
a. parietal peritoneum
b. greater omentum
c. lesser omentum
d. visceral peritoneum

Answer: D

The _____ lines the surfaces of the body wall.
a. mesentery
b. mesocolon
c. parietal peritoneum
d. visceral peritoneum

Answer: C

HISTOLOGY

Select the correct sequence beginning with the innermost layer and ending with the outermost layer.
a. Serosa, submucosa, muscularis externa, mucosa
b. Mucosa, submucosa, muscularis externa, serosa
c. Serosa, muscularis externa, submucosa, mucosa
d. Muscularis externa, serosa, mucosa, submucosa

Answer: B

This layer contains a plexus (plexus of Meissner) that has sensory neurons, parasympathetic ganglionic neurons and sympathetic postganglionic fibers.

a. Mucosa
b. Muscularis externa
c. Submucosa
d. Serosa

Answer: C

This layer is dominated by smooth muscle.
a. Serosa
b. Muscularis externa
c. Mucosa
d. Submucosa

Answer: B

MOVEMENT

The muscular layers of the GI tract are made up of:
a. visceral smooth muscle
b. cardiac muscle
c. skeletal muscle
d. parietal smooth muscle

Answer: A

_____ is defined as waves of muscular contractions that help move a small mass of digestive content throughout the GI tract.
a. Secretion
b. Peristalsis
c. Segmentation
d. Ingestion

Answer: B

The cycle of contraction that helps churn and fragment the bolus as well as mixes the contents with secretions of the intestine is called:
a. segmentation
b. peristalsis
c. emesis
d. defecation

Answer: A

CONTROL OF DIGESTION

The activities of the digestive systems are regulated by:

a. the nervous system
b. the endocrine system (hormones)
c. local mechanisms (local messengers)
d. A, B and C are all correct

Answer: D

Select the incorrect statement.
a. Short reflexes are also called myenteric reflexes.
b. Short reflexes control localized activities involving small segments of the digestive tract.
c. Long reflexes control large-scale waves of peristalsis that move materials from one region to another in the digestive tract.
d. The enteric nervous system is a neural network that coordinates the hormones along the digestive tract.

Answer: D

The digestive tract makes at least _____ hormones.
a. six
b. ten
c. twelve
d. eighteen

Answer: D

Hormones produced by cells in the digestive tract use the _____ to reach their target organs.
a. bloodstream
b. lymph
c. urine
d. saliva

Answer: A

Select the chemicals important in coordinating a response to changing conditions that affect a small portion of the digestive tract.
a. Histamine
b. Prostaglandins
c. Interleukins
d. A and B are correct
e. B and C are correct

Answer: D

ORGANS AND ACCESSORY ORGANS

The functions of the _____ are to provide sensory analysis, lubrication and mechanical processing of food materials.
a. tongue
b. salivary glands
c. teeth
d. cheeks
e. oral cavity

Answer: E

The pairs of salivary glands include _____.
a. sublingual
b. parotid
c. submandibular
d. submaxillary
e. A, B and C are all correct

Answer: E

Saliva is mostly made out of _____.
a. electrolytes
b. buffers
c. water
d. enzymes
e. antibodies

Answer: C

This enzymes present in saliva helps break down carbohydrates.
a. Salivary lipase
b. Salivary amylase
c. Trypsin
d. Peptidase

Answer: B

Which of the following pass through the pharynx?
a. Air
b. Food
c. Liquids
d. A, B and C are all correct

Answer: D

The process of chewing food is referred to as _____.
a. ingestion

b. digestion
c. mastication
d. egestion

Answer: C

Select the phases of swallowing.
a. Buccal, pharyngeal, esophageal
b. Pharyngeal, esophageal, gastric
c. Esophageal, gastric, intestinal
d. None of these choices

Answer: A

Select the functions of the stomach.
a. Produce intrinsic factor
b. Disrupt chemical bonds in food
c. Store ingested food
d. Mechanically breakdown ingested food
e. A, B, C and D are all correct

Answer: E

Select the term that does not belong.
a. Fundus
b. Cardia
c. Pylorus
d. Body
e. Esophagus

Answer: E

What is the function of HCl in the stomach?
a. The acidity of the gastric juice helps kill many microbes that are ingested with food.
b. The acidity denatures proteins and helps inactivates many enzymes in our food.
c. The acidity assists in the breakdown of plant cell walls and the connective tissue associated with meat.
d. The acidity helps activate pepsin (enzyme) which is needed for protein digestion.
e. All of these choices are correct

Answer: E

_____ % of nutrient absorption occurs in the _____.
a. 50; stomach
b. 90; small intestine
c. 75; large intestine

d. 33; mouth

Answer: B

The majority of chemical digestion and nutrient absorption occurs in the _____.
a. stomach
b. duodenum
c. jejunum
d. ileum

Answer: C

Select the correct sequence of the small intestine structures beginning with the segment that connects to the stomach.
a. Duodenum, ileum, jejunum, ileocecal valve
b. Ileum, jejunum, ileocecal valve, duodenum
c. Duodenum, jejunum, ileum, ileocecal valve
d. Ascending colon, transverse colon, descending colon, sigmoid colon, anus

Answer: C

Lymphatic capillaries are also known as:
a. blood capillaries
b. pulmonary capillaries
c. systemic capillaries
d. lacteals

Answer: D

Select the structures that increase the surface area of the small intestine.
a. Plicae circulares
b. Villi
c. Microvilli
d. A, B and C are all correct

Answer: D

Select the incorrect matching.
a. Pancreatic amylase; Enzyme breaks down starches
b. Pancreatic lipase; Enzyme breaks down complex lipids
c. Nucleases; Enzymes break down nuclear envelopes
d. Proteases; Enzymes break down large protein complexes
e. Peptidases; Enzymes break down small peptide chains

Answer: C

The functions of this organ are endocrine and exocrine.
a. Spleen
b. Liver
c. Kidneys
d. Pancreas

Answer: D

Select the functions of the liver.
a. Carbohydrate metabolism, lipid metabolism, amino acid metabolism
b. Waste product removal, vitamin storage, mineral storage, drug inactivation
c. Phagocytosis, synthesis of plasma proteins, removal of circulating hormones, removal of antibodies
d. Removal or storage of toxins, synthesis and secretion of bile
e. A, B, C and D are all correct

Answer: E

Select the mismatched pairing.
a. CCK; Stimulates production of pancreatic enzymes
b. Secretin; Stimulates production of alkaline buffers
c. Gastrin; Stimulates production of acids and enzymes
d. GIP; Dilate intestinal capillaries

Answer: D

Bile is synthesized in the _____ and stored in the _____.
a. liver; gallbladder
b. pancreas; liver
c. spleen; liver
d. None of these choices

Answer: A

Bile consists of _____.
a. water
b. ions
c. bilirubin and cholesterol
d. bile salts
e. A, B, C and D are all correct

Answer: E

The three parts of the large intestine are _____.
a. Ileum, cecum, colon
b. Cecum, colon, rectum

c. Haustra, taeniae coli, epiploic appendages
d. Cardia, fundus, body

Answer: B

Which of the following are produced by bacteria living in your colon?
a. Vitamin K and Vitamin B5
b. Biotin
c. Vitamin C and Vitamin E
d. Vitamin B12
e. A and B are correct

Answer: E

This pigment gives feces a brown color.
a. Carotene
b. Melanin
c. Keratin
d. Stercobilin

Answer: D

The colon can be divided into _____ regions.
a. two
b. three
c. four
d. six

Answer: C

Fecal material is:
a. 75 % water, 5 % bacteria and 20 % mix of indigestible materials, inorganic matter and epithelial cells
b. 5 % water, 75 % bacteria and 20 % mix of indigestible materials, inorganic matter and epithelial cells
c. 20 % water, 75 % bacteria and 5 % mix of indigestible materials, inorganic matter and epithelial cells
d. None of these choices

Answer: A

Made in the USA
Las Vegas, NV
24 April 2024

89112292R00017